MUSCLE & BONE

BY PAUL-VICTOR WINTERS

ACKNOWLEDGMENTS

Grateful acknowledgment is made to the following publications in which these poems, or earlier versions, have appeared: *Coastal Forest Review, Coracle Poetry, Footwork: The Paterson Literary Review, Journal of New Jersey Poets, New Jersey Review of Literature, Stockpot,* and *Without Halos.*

The author wishes to thank Stephen Dunn for his support and guidance.

Muscle & Bone
Copyright © 1995 by Paul-Victor Winters
All rights reserved

Design: James Laird
Editors: Margo Stever and Stephanie Strickland

Special thanks to Billy Collins who selected this chapbook as the winner of the 1995 Slapering Hol Press Chapbook Competition and to the following people who participated in the selection and production of this publication: Nicholas Carbo, Suzanne Cleary, Denise Duhamel, Jean Gonzales, Madeline Holzer, Barbara Nackman, Leslie Shipman.

ISBN 0-9624178-5-8

Slapering Hol Press
P.O. Box 366
Tarrytown, New York 10591

In memory of my mother

and for David

Contents

Eyes of God	7
First Poem after My Mother's Death	9
Poem on the Idea of Leaving	10
Widow	12
The Hyacinth	14
Widower	17
Poem in Three Parts for My Mother's Mother	20
Unrequited Love Poem	24
Hansel and Gretel	25
Scene: Northern Vermont, November	30

Eyes of God

Give the old man eyes, the woman says,
slicing carrots.
The boy pulls his lead pencil

out from between two pages
of his math book and slashes
two large X-marks

where the old man's eyes ought to be.
Outside, it is raining,
but gently. The woman next door

calls her boys inside.
He's dead, the boy says,
and his eyes are crossed out

like in cartoons.
Below the old man's peach face,
somewhere lost within the grey wind

of his beard, the boy scratches the word
God. Rain persists.
A good, strong wind moves the old poplar

tree. Rain hits the aluminum siding
of the old house. The old fingers
of the tree tap and slide

across the kitchen window.
The backyard football game is over.
The woman smokes another

of her husband's cigarettes,
pours juice into a glass for the boy.
Andrew says rain is angels pissing.

I don't think I believe in angels.
God's an old dead guy and all the angels
quit. The boy looks at his mother,

then out the window,
beyond the moving branches
of the poplar, toward the spot

where a football game
is not taking place.
The woman slices radishes

and cuts her finger.
She presses it
into the palm of her hand,

pulls her hand against herself.
She can't say anything
to her son and he is still

looking out through the window
and the potatoes
and corn are boiling.

*They all just quit
and they're going to parachute down
here and get jobs and buy houses.*

The branches still hit the window.
The woman rises, lets her hand go,
turns the flames off the corn

and potatoes, takes her son's hand.
She is squeezing too tightly.
She can't say anything,

but she pulls the boy behind her
and opens the back door
and walks through it.

The rain is almost violent now.
She walks with him
to the center of the yard.

She holds both his hands
and arches her head back
so the rain will hit her sharply

in the face.
Look up, she tells him,
look up.

First Poem after My Mother's Death

The glass bowl on the cluttered kitchen
table is so large, we could fit both
our hearts in it, my father and I. We could

stuff the whole world in there, maybe.
I thought, perhaps, we'd fill it
with yellow daffodils, but we've had enough

of flowers. There are things
we should put in the bowl but do not.
Instead, we open all the pill

bottles we can find throughout the house
and pour hundreds of pills into the bowl,
almost filling it. We are so happy to empty

them, we toss the lids over our shoulders
and drop the empty bottles to our feet.
There is a pill for everything, so many

shapes and sizes, such amazing color.
There is an entire landscape in our bowl, sky
and clouds and earth and water. The pills

are like tiny flower blossoms and we hate them.
We have done this together. I have always
wanted to do this. We have a large

bowl full of pills and, for all the love
we've ever known, can't even imagine
what to do with it.

Poem on the Idea of Leaving

Someday, I tell you,
   I'm going to leave.
There are moments, still,
when I reach
   for the old rosaries
and stop myself,

grieve every moment
   I've ever spent lost
in belief.
I'm going to leave,
   I tell you,
but the telephone rings,

the radio is loud,
   the house is full
of motion. If I let you,
you hold me, whisper
   or sing in my ear
and the sound is like wings

in motion and I want to get out
   of your hug, walk out
the door, fly
over the roofs of the houses
   across the street,
up over the trees

and skyline.
   I think about it at night
when talk in our bedroom
is strong.
   I am only ever thinking
about leaving.

You touch my shoulder blades.
   Just take me somewhere,
I say to God,
take me somewhere
   else and I'll stop lying,
but you are holding down

my shoulder blades
   and whispering something.
When you are asleep,
I fly around the bedroom
   for a while, down the hallway,
into the kitchen,

and by morning
   I'm in bed again.
You never know;
you're only whispering
   in your sleep
and I just can't listen.

Widow

In the morning
I walk to the ocean,
just down the block,

beyond the newspaper stand
and the house my sister-in-law
died in. I unfasten

a few buttons
on my blouse, only a few.
Looking into the sea,

I might imagine a pair
of soft, wide-open arms,
but do not. In India,

I've heard, watching
their husbands burn
on the tops of pyres,

widows are expected
to throw themselves
into the flames.

I could easily throw
myself into the ocean.
I bury my hands in sand.

Gulls keen.
If I were to imagine
a door on the grey horizon,

I might walk through it.
I watch clouds parade
over the water.

My last husband
did not beat me
but whispered to me

at night,
pressing himself
against my body.

I will wait for the cool
fingers of high tide
to inch around me

and imagine he is with me
and, for a while,
sing to the ocean.

The Hyacinth

for C. Salvatore
...and the lower lip kisses: that is what
the great poets call the Kiss Indirect.
          —from the Sanskrit of Vatsyana
(1st Century A.D., India) by Tambimuttu

Thank you for the hyacinth.
   It has died already,
      as though nothing beautiful

could ever survive my touch.
   The kitchen no longer smells of it.
      Its fragile shape

has become more fragile, brown.
   When my mother died,
      my father threw all her flowers

and plants into a heap
   in the backyard, beneath
      the apple tree, to rot.

Before she died,
   I'd walk around her house
      while she slept in her armchair

and look at them.
   I thought they'd help me
      write poems or help me think

of something brilliant to say
   to her. Before the hyacinth
      you gave me died,

I took it in its pot
   and held it in my hand.
      I wanted to touch it,

wanted to understand its shape.
   Tonight, I look into the night
      sky and don't understand

    those bodies, either.
        Maybe we write poems
            because we are trapped

    in the shells of our bodies,
        because there are too many shapes
            and bodies we can't touch.

    Maybe we shouldn't write poems.
        Maybe the shapes of things
            aren't so important. Yin

    might kiss the side of the body
        Yang, but they aren't real.
            Before my mother died,

    I'd walk around her house
        while she snored in her bedroom
            and try to listen to her

    flowers. They said
        only their own names
            and even that was just imagined.

    The lilac knew only
        the word lilac. The violet,
            violet. Before the hyacinth

    you gave me died,
        I held it close to my face.
            I wanted to kiss it indirectly,

    just to know it better.
        I couldn't handle its beautiful,
            awkward shape. Maybe I write poems

    because I can't handle beautiful things,
        because I cannot kiss
            anything beautiful directly.

    I never said anything brilliant
        to my mother, never heard words
            escape from the leafy fingers

of a flower, never saw the night
   sky clearly, the way an astronomer
      might. I will never leave

my body and it will not teach me
   anything that isn't obvious
      and brutally animal. Body, maybe,

is the only word the body knows.
   I will not write poems
      about hyacinths or daffodils,

will not leave the kitchen tonight.
   Before the hyacinth died,
      I thought I might learn

how to love, thought it might teach
   me something essential,
      thought it might, somehow,

save me. I thought it might
   make me feel secure,
      keep everything inside me

in balance, make me want to write
   love poems. I thought it might
      rustle its leaves and petals,

tell me secrets, show me
   its soft white body.
      Before it died,

I loved the flower.
   Perhaps I'll toss it,
      plastic pot and all,

into the backyard
   where it will rot. Perhaps
      I'll have gained nothing.

Widower

...he let the fog
swallow him, and his needs, and
his unexplainable things, whole.
—Michael Dalelio

Birds fly in groups
or alone.
I name them all.

I don't recognize them.
Their strange shapes
and messy wings

and shut-closed beaks
seem new to me
and I give them names:

Great Blue Soul-sucker,
Black-tipped Misanthrope,
Eastern White What-not.

They are sad. Their grace
is awful; their movements
and pitiful configurations,

the empty drone of the sound
of their moving,
don't fill me with anything.

Balls of clouds drift,
listless. They are going nowhere
and I am standing still

in marsh grass.
A good, thin fog
starts in at my ankles.

Birds sail by
and I let them go
nameless. I could lie

down in the fog
and marsh grass.
I could say her name,

say all her names.
I could offer her names
back to the sky,

say a name
as each ugly bird
passes.

I could buy myself a gun
and shoot down every damn
Soul-sucker that breezes by.

I'd say a name,
take a shot,
watch a bird fall

with each mention of her,
all the miserable birds
falling from the sky

into the restless fog
and cat-tails,
good and dead.

I could go back
to the house
we made together,

sleep in the bed
we made love in.
I could empty

the house of all
her things and fill
it with stuffed birds.

I could walk over to the bay
where the fog is thicker
and let it overtake me.

God damn the birds.
God damn the sky,
the clouds.

God damn the marsh grass
and everything in it.
God damn the fog.

God damn anything
she didn't love.
God damn everything

we never did.

Poem in Three Parts for My Mother's Mother

I.

Teacups drop onto the countertops
in our quiet kitchens.
You have seen too many people die.

You think you are the oldest person
in the world. Rain and hail
shake my bedroom window.

I wait for a telephone call
or a letter that will never arrive.
You let your African violets die.

You do not want to work at things.
Young men in kitchens write letters
they know they will never send.

Old women in kitchens choke
on their own words, try to swallow
conversations and give up,

leave the room, rasp the names
of the children they've lost.
We would stay put forever

if we thought it were possible,
if we knew our nightmares
would stop returning

every night, if the angels
of memory were more kind to us.
There are things we can't stop

seeing, can't rid ourselves of.
When your eyes are closed,
you see only the deep purple

and soft green skin
of your daughter's arms and neck,
thinly covering plastic tubes

full of liquids that will not save her,
her wide-open eyes, tubes, machines—
We are no longer able to speak

of the suffering. We pray
to our own dead;
there is little else.

II.

Imagine opening the front door
of a house at dusk in winter.
Let's say you've lived there

a long time,
that you're comfortable
there. Let's say the ring

of the doorbell surprises you.
Imagine opening the door.
It has just begun to snow.

Imagine everyone
you've ever loved
gathered in a large crowd.

We fill the whole yard.
We sing for you
all your favorite songs,

ignore the cold.
We sing for a long time
and the clear ring

of our voices,
rising together,
is almost as strong

as our love.
We are no longer
cold. You're not in that doorway

anymore; you don't know
where you are,
but you are comfortable.

We will sing forever.
There is nothing
but the sound of our voices.

III.

When I dream of you,
you're a child and I am your caregiver.
Shamrock plants and hyacinths

and African violets line the windowsill
in the kitchen of our house.
You are outside in the yard,

singing, playing games,
and I am at the kitchen table,
writing, or stroking a cat

and the sink is full
of dishes from a good meal
we've eaten together.

There is a pond in the backyard
and when the weather is nice,
I like to watch you

in your braids,
swimming around the shallow edges
of it. Last night,

I dreamt of a storm,
a deep grey sky, strong winds
tossing blades of rain

across the entire landscape.
You were in the yard
in your nightgown.

Thunder snapped
and lightning cracked
above you and you walked

into the awful water
of the pond, walked without
pausing, just walked into it

as though it called to you.
I ran outside, calling
your name, whatever name it was,

jumped into water,
felt around frantically
for your little body

and only grabbed water.
I finally found you
and pulled you up.

Leave me, you shouted
and the back door flew open
and closed, crashing over

and over in the powerful winds
of the storm. *Leave me*,
you yelled,

I want this.
There was nothing—
Jesus Christ—

nothing I could do
but let your small body go
and watch you sink,

slowly, your nightgown
billowing
into grey bloom.

Unrequited Love Poem

> Leaving, you smoothed your long white hair
> like a man who embraced his failures.
> —Tu Fu, "Dreaming of Li Po"

Some mornings,
I wake and find myself surprised
by the emptiness beside me.

I am always trying
to hold something
not there, always in love

with something
that does not exist.
Tonight, a winter storm

begins to swell
outside my bedroom window.
I love wind,

its strength and harshness.
I love the song
of loose windows rattling.

A light rain starts;
I love that, too.
I go to the bathroom

and look in the mirror.
I shave already grey hair
from my young face.

I may never know love,
may never get it right.
I may sleep forever,

dream of you while rain
hits the windowpanes,
and dawn slides into the room

from behind the blue
curtains; but I love you, partly,
because you are not here.

Hansel and Gretel

I. Finding the Candy House

After a while,
   the walk was no longer fun.
     Tree roots curled

at the children's feet
   like conquered snakes.
     The rise and gentle drop

of their feet made a music
   entirely new to them.
     Keep walking,

the boy told the girl,
   keep walking.
     He feared that they were lost

and feared that she'd say so.
   The bread crumbs falling
     behind the boy fell less often;

his bag became lighter.
   She heard black birds
     behind them, sweeping down

and swallowing the trail.
   She feared they'd walk
     around the entire round

world, but said nothing.
   Dusk fell. He turned
     and began retracing

their steps, lost.
   They walked for a long time,
     still not even close to home,

where soup was ready
   and their parents
     watched the day grow dark

from behind their patchwork
   curtains. The candy house,
     then, appeared before them

like a mirage on the forest's
   narrow horizon.
     He thought it looked more

like a good home than anything
   he'd ever seen,
     something like El Dorado.

She didn't trust it entirely,
   but went there. It was dusk;
     she was tired and cold.

He was blinded
   by the glimmer of peppermint;
     they'd already strayed too far.

II. Caged

For twelve days, the girl worked
   in the garden, cleaned house,
     did whatever the witch wanted.

The boy was caged in the yard,
   could not stand upright,
     crouched, became weak.

She thought he was pathetic
   and it was his fault anyway.
     She felt guilt, but at times,

carrying bushels of vegetables
   or baskets of dirty laundry,
     enjoyed seeing him there.

He kept quiet.
   He had become small
     and she was becoming large.

For twelve days,
   as she passed his cage,
      he'd said, *Sister,*

what are we going to do?
   and she wouldn't say anything.
      The witch planned on eating the boy.

The girl decided
   she wouldn't let it happen,
      that somehow she'd set her brother free.

She found a thin twig
   and gave it to him
      and, for twelve days, the blind witch

felt it through the thick
   bars of the cage, thought it
      the boy's thin finger

and kept feeding him
   chocolate and gingerbread.
      The girl had bought them time

and hoped to be home soon,
   hoped that when they made it,
      they'd still be children.

Still, she couldn't help
   but wonder what it might be like—
      the old woman's vicious teeth

tearing into the rough skin
   of the older brother
      she both hated and loved,

the witch's hands and mouth
   touching the new shape
      of his body,

cooked and ruined,
   the growing boy reduced
      to mere muscle and bone.

III. Witch-Burning

Between the pointed tongues
   of ardent flames, he sees her
      eyes, surprising himself.

There is much
   I could have learned,
      he thinks, from this woman.

He looks into the wretched face
   for a long while.
      Gretel is screaming, *Let's go!*

Let's go! He watches the witch burn,
   the flames still strong,
      the smell becoming awful.

He is surprised at himself,
   surprised by this victory,
      terribly ashamed.

He is watching her eyes.
   The moment before the thick
      lids close, her eyes become

motionless. She is staring
   at him. He notices something
      familiar in them. He sees in her

eyes something he's seen
   in the eyes of everyone
      he's ever loved.

He thinks he'll never love again.
   Her lips part slightly
      and she only mildly gasps.

She doesn't die, yet.
   She is still moving, slowly.
      She is almost beautiful.

He imagines she is speaking to him:
My boy, it is all right.
Be calm. Go home now.

There is no such thing as a mistake.
No one here has done anything wrong.
Our story is over. Go, now.

Gretel is still screaming
from the doorway, waving her arms.
He wishes to hold the old woman.

He may never hold another
with love again. The moment
before the sad shapes

of her body snap and collapse,
browned, he wishes to kiss her—
a child's kiss, a kiss to send her

off with, one which might, somehow,
save her. He cannot, of course,
and the boy leaves, grown.

Scene: Northern Vermont, November

With two
hands,
you snapped
a large branch
and tossed
the pieces
aside.
Keep walking,
you said.
Our boots fell
into the earth's
frozen layer
of snow,
making noise.
I had trouble
keeping
up with you.
It was dusk.
We found
our lean-to.
I made a pit
and you cut
wood.
We made fire.
Our bodies
filled
the bag
the way black
fills
the sky.
I looked up
at the stars,
little fires
in the blackness,
little fires
above us.

THE HUDSON VALLEY
WRITERS' CENTER

The Hudson Valley Writers' Center aims to nurture the remarkable literary heritage of our region and to introduce people to the power of the literary arts. Under the imprint of Slapering Hol (Old Dutch for Sleepy Hollow) Press, The Center publishes an annual volume of poetry, alternating thematic anthologies with chapbooks by poets who have not previously published in book form. The Center's reading series provides opportunities for both emerging and established writers to be heard. Fall and spring workshops in poetry, essay, fiction, play, travel, and screenwriting are offered, and outreach programs send trained writers to lead workshops for those with special needs, including homeless and institutionalized children. The Center has also initiated the restoration of the abandoned, but historically significant, railroad station at Philipse Manor, North Tarrytown, New York, for use as audience, workshop, and office space.

SLAPERING HOL PRESS

What's Become of Eden: Poems of Family at Century's End (1994)

Chapbook Series

Muscle & Bone   Paul-Victor Winters (1995)

Muscle & Bone is a mesmerizing first collection of poems spoken in a voice at once alert and dreamy, nervy and vulnerable, tentative and flighty. On the page, the poems are lean and plainly spoken, but they are alive with surprises and bright maneuvers. The result is a fine combination of deftness of craft and ease of expression.   —Billy Collins, Judge

Weathering   Pearl Karrer (1993)

River Poems: An Anthology (1992)

Note for a Missing Friend   Dina Ben-Lev (1991)

Voices from the River: An Anthology (1990)

The Hudson Valley Writers' Center, Publishers